Ecosystems Research Journal

Everglades Research Journal

Robin Johnson

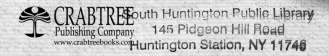

CRABTREE
Publishing Company
www.crabtreebooks.com

Crabtree Publishing Company
www.crabtreebooks.com

Author: Robin Johnson

Editors: Sonya Newland, Kathy Middleton

Design: Clare Nicholas

Cover design: Margaret Amy Salter

Proofreader: Angela Kaelberer

Production coordinator and prepress technician: Margaret Amy Salter

Print coordinator: Margaret Amy Salter

Consultant:

Written and produced for Crabtree Publishing Company by White-Thomson Publishing

Front Cover:

Title Page:

Photo Credits:

Cover: iStock: title page (snake) SouthernBelladonna
All other images from Shutterstock

Interior: Alamy: p. 9 Ingolf Pompe 9, p. 10b Jim West, p. 13b Susanne Masters, p. 15b National Geographic Creative, p. 23l EPA, p. 25t Bob Gibbons; Ron Dixon: pp. 4, 28; iStock: p. 15t IMPALASTOCK, p. 15m SouthernBelladonna, pp. 28–29 ablokhin; Shutterstock: p. 4b gusenych, pp. 4-5 pisaphotography, p. 5 matthieu Gallet, p. 6t Christopher Lance, p. 6b tome213, p. 7t Beth Swanson, p. 7b Andrew M. Allport, p. 8 timyee, pp. 8–9 Rudy Umans, p. 10t William Silver, p. 11t Beth Swanson, p. 11b feel4nature, p. 12t mariakraynova, p. 12m Ondrej Prosicky, p. 12b Don Mammoser, p. 13t Daniel Zuppinger, p. 14 Chiyacat, p. 16t Tory Kallman, p. 16b Chesapeake Images, p. 17tl IrinaK, p. 17r IrinaK, p. 17b Rich Carey, pp. 18–19 Thannithi, p. 9t Astrid Gast, p. 19m Macrovector, p. 19b THPStock, p. 20 Perry Correll, p. 21t Heiko Kiera, p. 22m Dennis Kartenkaemper, p. 22b Samuel Acosta, p. 22t jo Crebbin, p. 22bl j Loveland, p. 22br FloridaStock, p. 23r jeff gynane, p. 24t William Silver, p. 24b Steve Bower, p. 25b Karuna Eberl, p. 26 Tania Thomson, p. 27t Matt Jeppson, p. 27b John A. Anderson; Wikimedia: p. 29 Ebyabe.

Library and Archives Canada Cataloguing in Publication

CIP available at the Library and Archives Canada

Library of Congress Cataloging-in-Publication Data

Names: Johnson, Robin (Robin R.), author.
Title: Everglades research journal / Robin Johnson.
Description: New York, New York : Crabtree Publishing Company, 2018. |
Series: Ecosystems research journal | Includes index.
Identifiers: LCCN 2017029504 (print) | LCCN 2017030906 (ebook)
  ISBN 9781427119292 (Electronic HTML) |
  ISBN 9780778734697 (reinforced library binding : alkaline paper) |
  ISBN 9780778734949 (paperback : alkaline paper)
Subjects: LCSH: Everglades (Fla.)--Environmental conditions--Research--Juvenile literature. | Biotic communities--Florida--Everglades--Juvenile literature. | Ecology--Research--Florida--Everglades--Juvenile literature. | Everglades (Fla.)--Description and travel--Juvenile literature.
Classification: LCC GE155.E84 (ebook) |
  LCC GE155.E84 J64 2018 (print) | DDC 577.6809759/39--dc23
LC record available at https://lccn.loc.gov/2017029504

Crabtree Publishing Company
www.crabtreebooks.com   1-800-387-7650

Printed in Canada/082017/EF20170629

Copyright © **2018 CRABTREE PUBLISHING COMPANY.** All rights reserved. No part of this publication may be reproduced, stored in a retrieval system or be transmitted in any form or by any means, electronic, mechanical, photocopying, recording, or otherwise, without the prior written permission of Crabtree Publishing Company. In Canada: We acknowledge the financial support of the Government of Canada through the Canada Book Fund for our publishing activities.

Published in Canada
Crabtree Publishing
616 Welland Ave.
St. Catharines, Ontario
L2M 5V6

Published in the United States
Crabtree Publishing
PMB 59051
350 Fifth Avenue, 59th Floor
New York, New York 10118

Published in the United Kingdom
Crabtree Publishing
Maritime House
Basin Road North, Hove
BN41 1WR

Published in Australia
Crabtree Publishing
3 Charles Street
Coburg North
VIC, 3058

Contents

Mission to the Everglades	4
Field Journal Day 1: Lake Okeechobee	6
Field Journal Day 2: Shark Valley	8
Field Journal Day 3: Florida Bay	10
Field Journal Day 4: Flamingo to Pearl Bay via Coot Bay	12
Field Journal Day 5: Pearl Bay to Shark River	14
Field Journal Day 6: Shark River to Harney River via Graveyard Creek	16
Field Journal Day 7: Harney River to Broad River via Nightmare	18
Field Journal Day 8: Broad River to Willy Willy	20
Field Journal Day 9: Willy Willy to Plate Creek	22
Field Journal Day 10: Plate Creek to Sunday Bay	24
Field Journal Day 11: Sunday Bay to Everglades City	26
Final Report	28
Your Turn	30
Learning More	31
Glossary & Index	32

Mission to the Everglades

Time to get out my rubber boots! I just found out I am going on a research trip to the Everglades in Florida. I have wanted to explore this huge wetland area ever since I became a **wildlife biologist**. A wetland is land covered with water for part or all of the year. The Humans for Habitats organization is sending me there. They want me to see how a decrease in fresh water is affecting the land, plants, and animals. I will also look for species that do not belong there, signs of pollution, and other threats. My report will include the actions being taken to save the Everglades.

I started planning my trip right away. I will spend most of my time in Everglades National Park. It is one of the largest wilderness areas in the United States. There are two seasons in the Everglades: dry season and wet season. The wet season is too hot and rainy to observe the **ecosystem** easily. I have decided to travel during the dry season when water levels are low. Animals will have to gather together at water holes to drink. That is perfect for curious scientists!

There are thousands of different plants and animals in the Everglades. These include more than 360 types of birds.

Dry season lasts from about November to March. Wet season lasts from about April to October.

The Everglades used to cover almost 11,000 square miles (28,000 square kilometers). People began to drain water from the land in the late 1800s so they could farm and build towns and roads. Only half of the original wetlands are left today.

Field Journal: Day 1

← Spatterdock.

↑ Lake Okeechobee used to overflow its banks and create a shallow river that flowed all the way down to Florida Bay.

Lake Okeechobee

I started my research trip where the Everglades ecosystem begins, at Lake Okeechobee. I climbed into a canoe and set out to explore Florida's largest lake. I was surprised how shallow the water was. I could touch the bottom with my paddle in some places! I did not want to disturb the delicate plant life, though. The lake was dotted with water lilies, American lotus, and spatterdock. Plants that live on or near water are called **aquatic** plants. They provide shade, shelter, and food for fish, birds, and other animals.

I spotted some ring-neck ducks nibbling on water lilies, and large-mouth bass swimming near the canoe. I was hoping to see a snail kite hunting along the shore. These are rare birds, and they are getting harder and harder to find. Snail kites and other animals lost much of their wetland habitat when water was drained from the Everglades.

The Everglades receive only one-third the amount of fresh water they used to get from the lake.

1/3

The large-mouth bass is one of 300 fish species found in the Everglades.

natstat STATUS REPORT ST456/part B

Name: Snail kite (Rostrhamus sociabilis)

Description:
Snail kites are **birds of prey** that eat mainly apple snails. The birds use their thin, curved beaks to pull the snails out of their shells. Apple snails are getting hard to find, however. They live in freshwater wetlands. Less fresh water in the Everglades means less prey for snail kites.

Threats:
Habitat loss.

Numbers:
400 breeding pairs.

Status:
Endangered.

Attach photograph here

Field Journal: Day 2

Shark Valley

Today I traveled by car into Everglades National Park. This is the heart of the freshwater **marsh**. I was excited to hike in the Shark River Slough. A slough is a low area of land that the water follows through the Everglades. There are no sharks this far from the ocean, but I did see dozens of American alligators! The gators were swimming and sunning themselves on the banks of the river. I admired the huge reptiles but kept my distance from them. They usually eat small animals, but these **apex predators** can catch and eat large mammals too. See you later, alligator!

American alligators are important to their ecosystem because they keep the numbers of the animals they eat under control.

I climbed the observation tower to get a bird's-eye view of the wetlands. There were fields of sawgrass as far as the eye could see! This tall, grass-like plant is the most common plant in the Everglades. It grows in freshwater marshes that are too wet for most other plants to survive. A park ranger explained that even these hardy plants are in trouble. Rising sea levels are bringing more salt water into the ecosystem. Sawgrass grows in **peat**, which is a rich, mucky soil. Salt water is breaking down the peat and threatening the growth of sawgrass.

observation tower →

In the past 100 years, the sea level has risen by about 8 inches (20 centimeters) due to **climate change**. Pollution and other factors are making the temperature of air and water higher. This causes huge blocks and sheets of ice to melt into the ocean. The extra water slowly raises the sea level over time.

Endless sawgrass marshes give the Everglades their nickname, "River of Grass."

Field Journal: Day 3

Florida Bay

Today I explored Florida Bay at the southern edge of the Everglades. This is where the river meets the ocean. The water here is **brackish**, which is a mixture of fresh water from Lake Okeechobee and salt water from the ocean. I traveled by canoe because I did not want to disturb the delicate ecosystem with a motor boat. The water is shallow, and the propeller blades on motors can kill or hurt animals and tear up wetland plants.

I waved to a park ranger and some tourists who paddled quietly by in canoes.

About a million people come to the Everglades each year. Visiting an ecosystem can help people learn how to protect it. But it can also harm the plants and animals that live there.

Seagrasses provide food and shelter for fish, crabs, manatees, sea turtles, dolphins, and many other **marine** animals.

As I paddled around the bay, I was lucky to spot a Florida manatee in the water. This large "sea cow" was grazing on seagrasses. Seagrasses are plants with long, grass-like leaves that grow entirely underwater. Thick, green seagrasses swayed gently around the big manatee and the canoe. I noticed patches of dead brown seagrasses here and there, however. Without enough fresh water from Lake Okeechobee, the wetlands are becoming too salty for some species to survive.

natstat STATUS REPORT ST456/part B

Name: Florida manatee (Trichechus manatus latirostris)

Description:
Manatees are big, gentle, slow-moving marine **herbivores**. These "sea cows" help the ecosystem by grazing on seagrasses, which must be constantly cut short to grow and stay healthy.

Attach photograph here ➡

Threats:
Loss of water habitat and collisions with boats.

Numbers:
Less than 2,500 mature adults.

Status:
Endangered.

Field Journal: Day 4

Flamingo to Pearl Bay via Coot Bay

This morning I met my wilderness guide at the town of Flamingo. She will steer me through the maze of rivers, creeks, and ponds that make up the Everglades. We loaded up our gear and began our eight-day canoe trip north. The Sun was shining and I was glad I had planned my research trip for the dry season. I would avoid the storms of the wet season, and the frequent fires caused by lightning striking trees and other plants.

Sightings

I spotted a grasshopper on a burned tree. This was a sign there had been a fire here. Natural fires are helpful to the ecosystem. They burn off dead leaves and plants and allow new ones to grow in their place. The roots of many wetland plants grow underwater and are safe from fire.

As we paddled slowly through the wetlands, I saw birds of all kinds flying or wading in the water. Some species nest here all year round. Others **migrate** here during the dry season. The creeks and ponds are shallow now, so it is easier for birds to find fish and other prey. I observed a great blue heron fishing at an alligator hole—but not for long! It quickly flew away when it spotted the crafty gator that had made the hole.

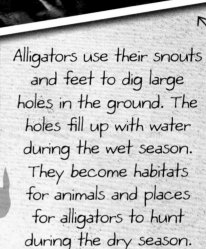

Alligators use their snouts and feet to dig large holes in the ground. The holes fill up with water during the wet season. They become habitats for animals and places for alligators to hunt during the dry season.

We were ready to set up camp after a full day of paddling. We pitched our tent on a raised wooden platform called a chickee. These platforms are built for campers where there is no dry land.

Other bird species I spotted included white ibis, wood stork, roseate spoonbill, and anhinga.

There are 90 percent fewer wading birds in the Everglades now than in the past. Millions of birds were killed for their feathers in the 1800s. They were used to decorate women's hats.

Field Journal: Day 5

Pearl Bay to Shark River

We continued paddling north through the twisting waterways. We stayed near the shore but were careful to not disturb the red mangroves along our path. Mangroves are incredibly important trees in the Everglades ecosystem. Storms in the wet season cause **erosion** and carry away soil that plants need. The strong, tangled roots of the mangrove trees grow above the ground and help hold the soil in place. We found it hard to paddle around the big roots that reach out into the narrow creeks. We barely made it through some tight spots!

More than one-third of the world's mangrove trees have disappeared in the last 100 years.

I observed many kinds of animals as we paddled slowly through the mangrove **swamp**. Wading birds were feeding and nesting in the branches of the mangroves. Fish, crabs, and other animals were hiding in the sturdy, twisted roots of the trees in the water. Mangroves act as **nursery habitats** for many young animals. As the animals grow, the trees provide them with food and shelter from predators. Sadly, mangrove forests are in danger. They grow in brackish swamps. The water here may become too salty for them to survive. Rising sea levels also threaten to drown these important trees.

Alligators, snakes, turtles, and many other reptiles find food and shelter among the mangroves.

natstat STATUS REPORT ST456/part B

Name: Atlantic goliath grouper (*Epinephelus itajara*)

Threats: Loss of mangrove habitat.

Description: Atlantic goliath groupers are saltwater fish that live in mangrove swamps. They move to the ocean when they are large enough to survive there. These fish can grow to more than 8 feet (2.5 meters) long! Their huge size made them a popular fish to catch in the past. It is now against the law to harm this fish species.

Numbers: Unknown.

Status: Critically endangered.

Attach photograph here

Field Journal: Day 6

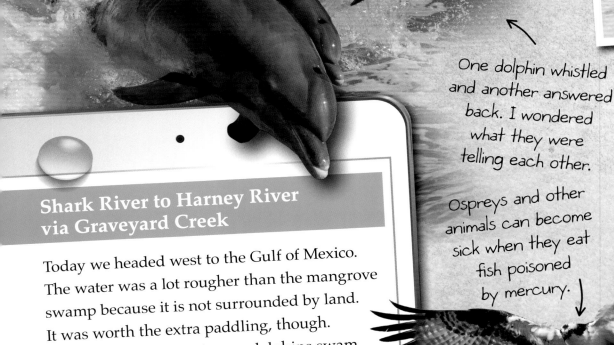

One dolphin whistled and another answered back. I wondered what they were telling each other.

Ospreys and other animals can become sick when they eat fish poisoned by mercury.

Shark River to Harney River via Graveyard Creek

Today we headed west to the Gulf of Mexico. The water was a lot rougher than the mangrove swamp because it is not surrounded by land. It was worth the extra paddling, though. A pod of Atlantic bottlenose dolphins swam right beside the canoe! I was glad to see them. High levels of **mercury** have been discovered in bottlenose dolphins off the coast of the Everglades. This poison is released by some power plants and factories. It pollutes the ocean and can make marine animals sick.

Sightings

I spotted a North American osprey flying high above the ocean. I watched this large bird of prey dive feetfirst into the water. It caught a slippery fish with its sharp talons.

Baby leatherback turtles crawl to the sea after they hatch. They must move quickly or risk being eaten by predators.

I set out to explore the beach on foot and found a wide trail in the sand. It seemed to lead from the water up to the beach. I studied the trail and identified it as a leatherback sea turtle crawl. Adult female leatherback turtles crawl out of the ocean to lay their eggs on beaches. I followed the tracks to a nest, but I was careful not to disturb it. There were about 100 eggs in a hole loosely covered with sand. It was hard to believe that the reptiles inside these little eggs would grow to become some of the biggest turtles in the world!

natstat STATUS REPORT ST456/part B

Name: Sea turtles

Description: Sea turtles face many threats besides being eaten by predators. Humans have endangered these reptiles the most. There are fewer places for sea turtles to lay their eggs because people have built homes and hotels along coasts. Rising sea levels have also reduced the size of the beaches where they nest. People have polluted the oceans where sea turtles live.

Threats: Habitat loss and pollution.

Numbers: Unknown.

Status: Endangered or threatened in the United States.

Attach photograph here

Field Journal: Day 7

Harney River to Broad River via Nightmare

Today we paddled through a narrow waterway called "Nightmare." It was a race against time to get here. This part of our route can only be crossed during a high tide when the water is at its highest level. The tides rise and fall daily in areas of the Everglades near the ocean. We did not want to get stuck in the mud at low tide! But the mangroves are thick here, and we could only inch our way slowly through them.

Tides are getting higher in the Everglades because the sea level is rising. Higher tides bring more salt water into the ecosystem, which may harm species in freshwater and brackish habitats.

I swatted mosquitoes as we paddled slowly through the swamp. They were annoying, but not nearly as bad now as they would be in summer. The Everglades buzz with billions of mosquitoes during the wet season. These insects can be pests—some even bite alligators! But they are an important part of wetland food chains. Fish, frogs, birds, bats, and many other animals eat mosquitoes.

Mosquitoes lay their eggs in the still water of the wetlands.

Sightings

I saw a mangrove crab scrambling quickly up a tree. These little crabs climb mangrove trees during high tide and return to the ground to look for food at low tide. They help the ecosystem by eating dead mangrove leaves that have fallen to the ground.

Field Journal: Day 8

Broad River to Willy Willy

Last night we set up our tent at a campsite at Broad River. It felt good to be back on solid ground after sleeping on a chickee. There were plenty of trees and other plants around us. One tree caught my eye, and I walked over to take a closer look. It was a Brazilian pepper. This is an **invasive species**, which means it does not belong in the Everglades. This bushy tree is native to South America. Brazilian peppers spread quickly and form thick forests. They take up space, block sunlight, and steal resources from plants that grow here naturally.

Brazilian peppers were brought to Florida in the 1800s for use as decorative plants. It is now illegal to sell or plant these trees anywhere in the state.

Invasive species compete with native plants and animals for food, water, sunlight, and other resources. Most invasive species have no natural predators in their new habitats.

We packed up and continued on our way until something stopped us in our tracks. It was a Burmese python attacking an alligator! These huge snakes are the biggest, strongest invaders in the wetlands. They are native to Southeast Asia but were released into the Everglades years ago. Scientists think the snakes were pets that grew too big for people to keep. Now there are thousands of Burmese pythons here. These snakes are powerful hunters that squeeze their prey to death.

Burmese pythons compete with alligators and other apex predators for food in the Everglades.

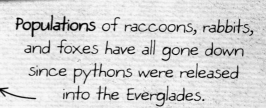

Populations of raccoons, rabbits, and foxes have all gone down since pythons were released into the Everglades.

Scientists have tracked and removed more than 2,000 Burmese pythons from the Everglades. It is now illegal to bring these deadly snakes into the United States.

21

Field Journal: Day 9

Willy Willy to Plate Creek

This morning I woke up early and went for a hike. I walked deep into the woods and was surprised to see a group of men there. I was even more surprised when I saw their rifles! These men were **poachers**, who were hunting illegally in the Everglades. I wondered if they were looking for Florida panthers, white-tailed deer, or other animals. I did not hang around to find out. We quickly packed up camp and jumped into the canoe. I texted the park ranger and reported their location as soon as I had cell-phone service.

Florida panthers are endangered due to habitat loss and poaching. Scientists believe there are fewer than 100 of these big cats left in south Florida.

Poachers do not hunt just animals in the Everglades. They also take ferns, orchids, and other rare plants, even though it is against the law.

I was afraid I might see more poachers as we wound our way north through the mangroves. Instead, I saw a group of researchers from the South Florida Natural Resources Center. We paddled up to them, and they explained that they were **monitoring** the water in the wetlands. They recorded the water level and temperature. Then they got a water sample to take back to their lab. They will measure how much salt is in the water and check it for signs of pollution. The researchers study this information to learn how human activity is affecting the ecosystem. They try to understand how the ecosystem is changing so they can figure out how to protect it.

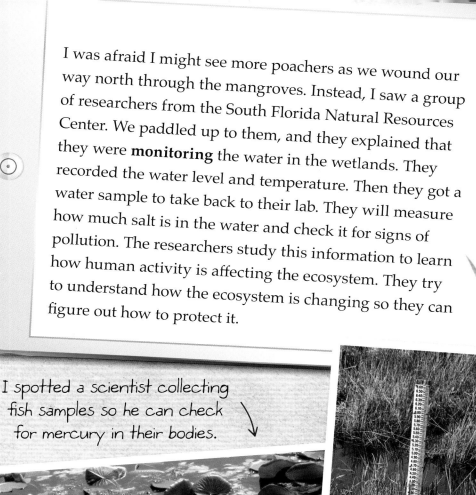

This gauge helps researchers measure water levels in the Everglades.

I spotted a scientist collecting fish samples so he can check for mercury in their bodies.

Field Journal: Day 10

Plate Creek to Sunday Bay

Today I cooled off in a hardwood hammock. It was not the type of hammock you might think though! A hammock is an area of high, dry land with tall, leafy trees. It is like a little tree island in the middle of the wetlands. Palm, gumbo-limbo, mahogany, and other hardwood trees were growing close together. They formed a thick canopy high above us. I looked up at the branches and saw orchids and other **epiphytes** growing there. Epiphytes are plants that grow on other plants. They get the water and nutrients they need from the air and rain.

A hardwood hammock.

I spotted a Cuban tree frog on a branch. This big tree frog is an invasive species that preys on insects, snails, lizards, snakes, and other frogs.

I almost fell into a solution hole at my feet as I was peering up at the trees! Solution holes are deep pits that are formed when rock is eroded over time. The holes are hard to spot because they fill with water during the rainy wet season. They often stay flooded all year round. Solution holes provide freshwater habitats for fish, alligators, and other animals in the dry season.

Solution holes help protect hardwood hammocks from fires during the wet season.

This part of my trip took me near Big Cypress National Preserve. Cypress forests grow in the Preserve and in some parts of the Everglades. Bald cypress are tall, leafy trees that can grow in **waterlogged** soil.

natstat STATUS REPORT ST456/part B

Name: Orchids

Description:
Orchids are plants with colorful, fragrant flowers. Rare orchid species are very valuable to collectors. People take them from the Everglades even though it is now illegal. Sadly, this has caused at least three orchid species in the ecosystem to become extinct.

Threats:
Poaching and habitat loss.

Numbers:
Unknown.

Status:
Many species are endangered.

Attach photograph here →

Field Journal: Day 11

Sunday Bay to Everglades City

By the last day of my trip I still had not seen a crocodile. I had almost given up when I spotted a big reptile lying in the mud with its mouth wide open. I got a good look at the animal's strong jaws and sharp teeth with my binoculars. I was able to identify it as an American crocodile because it had a long, pointed snout and gray-green back and tail. Alligators have wider, rounded noses and darker skin. Alligators live in fresh water, and crocodiles prefer brackish or salt water.

Crocodiles do not sweat. They open their mouths to release body heat and cool off. They also cool their bodies by resting in the shade and swimming.

The Everglades are the only place on Earth where crocodiles and alligators live side by side in the wild.

Not all reptiles in the Everglades are huge, though. Later I spotted a tiny green anole running up a tree. The little lizard was brown on the bark of the tree. Then it turned bright green while it rested on the leaf of the tree! The green anole uses camouflage to help hide from predators. It has **adapted** over time to survive in its habitat, like most species in the Everglades. But today there are new threats facing plants and animals in this ecosystem. Some may not survive.

Green anole.

Sightings

I was lucky enough to see a golden orb spider spinning a shiny, gold-colored web. The color of the web helps it blend into the sunlight. It was only by chance that I spotted it!

Final Report

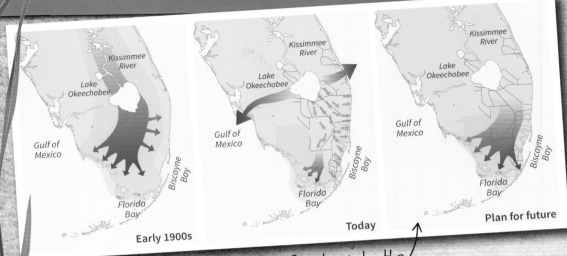

This diagram shows the flow of water into the Everglades over the past 100 years.

Report to: HUMANS FOR HABITATS

OBSERVATIONS

My research trip has made one thing crystal clear to me—the Everglades need fresh water. Different plant and animal species have adapted to live in fresh water, brackish water, or salt water. But rising sea levels caused by climate change, among other causes, are affecting the waters in the Everglades.

FUTURE CONCERNS

The wetlands are getting less fresh water and more salt water than ever before. Salt is destroying the peat and shrinking natural habitats. Invasive animals and plants are thriving, but many native species are in trouble.

Conservation Projects

Scientists are working hard to save the Everglades, however. The U.S. government and Florida have agreed to major plans to bring the ecosystem back to health. The goals are to bring fresh water back to the wetlands and decrease pollution. One important project involves building bridges along a highway that crosses through the Everglades. The road blocks the natural flow of water from Lake Okeechobee to Florida Bay. Bridges will allow fresh water to flow under the highway and through the Everglades again for the first time in nearly a century.

Eight million people rely on the Everglades for clean drinking water. Fresh water in the ecosystem soaks slowly into the ground. It is stored in underground caves called aquifers. The water is pumped out and sent to homes all around Florida. People would not be able to drink the water if too much salt water soaked into the aquifers.

← These tourists are exploring the Everglades so they can learn how to help save the ecosystem.

Your Turn

★ This journal is a work of fiction. It is based on research and real information, however. Look through the book to find maps, diagrams, and pictures that help support the story being told. What other types of sources could you add to tell the story? Write them down or draw them.

★ Journals allow people to share information and describe events from their own point of view. How would the Everglades journal be different if it were written by a student, a person from another country, or even an alligator? Get creative and write about events in the journal from a different point of view.

★ The Everglades and many other ecosystems around the world need fresh water. Keep track of all the ways you conserve—or waste—water throughout the day. What did you observe? How could you save more water? Write a journal entry about it.

Learning More

BOOKS

The Everglades by Katie Marsico (Cherry Lake Publishing, 2013)

Everglades National Park by Maddie Spalding (Core Library, 2016)

Preserving America: Everglades National Park by Nate Frisch (Creative Paperbacks, 2014)

The Wetlands of Florida by Peggy Sias Lantz and Wendy A. Hale (Pineapple Press, 2014)

WEBSITES

www.nps.gov/ever/learn/kidsyouth/learning-about-the-everglades.htm
Learn about water, wildlife, habitats, seasons, and more at the Everglades National Park's website.

www.dep.state.fl.us/secretary/kids/default.htm
You'll find activities, quizzes, fast facts, pioneers, and postcards on the kids' page of the Florida Department of Environmental Protection website.

www.pbslearningmedia.org/resource/ess05.sci.ess.watcyc.everglades/an-everglades-visit/
Visit the PBS Learning website to take a virtual tour of the Everglades with a young explorer.

Glossary & Index

adapted changed over time to become better suited to an ecosystem

apex predator an animal at the top of the food chain that has no natural predators

aquatic living in or near water

bird of prey a bird that hunts and eats meat

brackish slightly salty, as in a mixture of fresh water and salt water

climate change a change in the normal weather in an area over time that is caused by pollution and other human actions

ecosystem a community of plants, animals, and their environment

epiphytes "air plants" that grow on trees or other plants without harming them

erosion a process in which soil and rocks are worn away by wind and water over time

herbivore an animal that eats mainly plants

invasive species a plant or animal that is not native to an ecosystem and which can harm it

marine living in or near the ocean

marsh an area in a wetland without trees

mercury a poisonous metal that is sometimes released into the air and water from factories and power plants

migrate to move to a new habitat for a period of time, usually when the seasons change

monitor to observe and check the progress or quality of something over a period of time

nursery habitat a place where young fish, crabs, and other marine animals live

peat a type of soil made up of waterlogged, partly decayed plant materials

poacher someone who takes or kills wild species illegally

population the total number of a species living in an area

swamp an area in a wetland that has trees

waterlogged soaked or filled with water

wildlife biologist a scientist who studies animals and other wildlife in their ecosystems

adaptations 27, 28
alligator 8, 13, 15, 19, 21, 25, 26
apex predator 8, 21
Atlantic goliath grouper 15

Big Cypress National Preserve 4, 24
birds 5, 6, 7, 13, 15, 16, 19
brackish water 10, 15, 18, 26, 28
Brazilian pepper 20
Broad River 18, 20
Burmese python 21

chickee 13, 20
climate change 9, 28
crab 11, 15, 19
crocodile 26

dolphin 11, 16
dry season 5, 12, 13, 25

ecosystem 5, 6, 8, 9, 10, 11, 12, 14, 18, 19, 23, 25, 27, 29
epiphyte 24
erosion 14, 25

fire 12
fish 6, 7, 11, 15, 16, 19, 23, 25
Flamingo (town) 12
Florida Bay 4, 6, 10, 28, 29
Florida panther 22
fresh water 4, 7, 8, 9, 10, 11, 18, 25, 26, 28, 29
frog 19, 24

golden orb spider 27
grasshopper 12
Gulf of Mexico 4, 16, 38

hardwood hammock 24, 25
hunting 7, 22

invasive species 20, 28

Lake Okeechobee 4, 6, 7, 10, 11, 28, 29
large-mouth bass 7
lizard 24, 27

manatee 11
mangrove 14, 15, 16, 18, 19, 23
marsh 8, 9

mercury 16, 23
mosquito 19
nursery habitat 15

observation tower 9
orchid 22, 24, 25
osprey 16

peat 9, 28
plants 6, 9, 10, 11, 14, 20, 22, 27, 28
poaching 22, 23, 25
pollution 4, 17, 23, 29

salt water 9, 10, 15, 18, 23, 26, 28
sawgrass 9
sea level 9, 15, 17, 18, 28
sea turtle 11, 17
seagrasses 11

snail kite 7
snake 15, 21, 24
soil 9, 14, 25
solution hole 25
South Florida Natural Resources Center 23
spatterdock 6
storm 12, 14
swamp 15, 16, 19

tides 18, 19

water holes 5
water lilies 6, 7
wet season 5, 12, 13, 14, 19, 25
wetlands 4, 5, 7, 9, 10, 11, 12, 13, 19, 21, 23, 24, 28, 29

RECEIVED DEC 1 2017